HARCOURT Math FLORIDA

Fast Track to FCAT

Grade 1

Orlando   Austin   Chicago   New York   Toronto   London   San Diego

Visit *The Learning Site!*
www.harcourtschool.com

Grateful acknowledgment is made to the Florida Department of Education, Office of
Assessment and School Performance, Tallahassee, Florida, 32399-0400, for permission to
reprint FCAT Think, Solve, Explain logo and extended- and short-response scoring rubrics.

Printed in the United States of America

ISBN 0-15-339053-0

13 14 073 10 09

Contents

Problem-Solving Think Along

Understand

1. What is the problem about?

_____

2. What is the question?

_____

3. What information is given in the problem?

_____

Plan

4. What problem-solving strategies might I try to solve the problem?

_____

5. What is my estimated answer?

_____

Solve

6. How can I solve the problem?

_____

Check

7. How do I know whether my answer is reasonable?

_____

Taking Tests

Taking a test is one way to show what you know. Each time that you take a test, remember to:

- Listen to the teacher's directions.
- Read carefully.
- Mark the answers clearly.
- Raise your hand if you have a question.
- Work carefully.
- Do the best that you can.

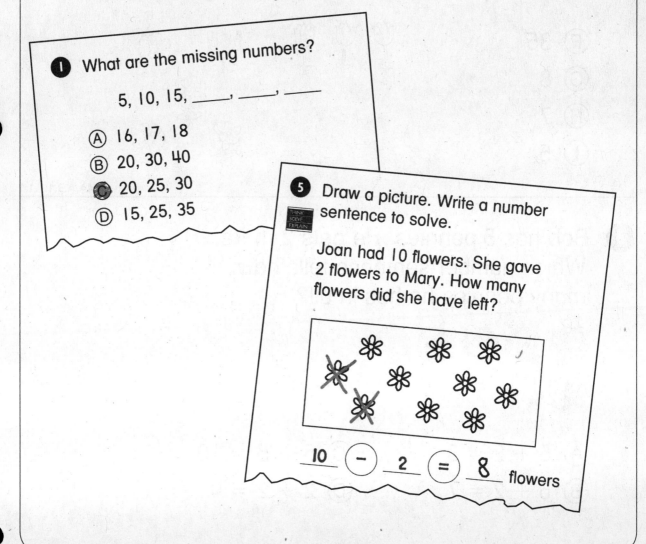

1 What are the missing numbers?

5, 10, 15, _____, _____, _____

- (A) 16, 17, 18
- (B) 20, 30, 40
- (C) 20, 25, 30
- (D) 15, 25, 35

5 Draw a picture. Write a number sentence to solve.

Joan had 10 flowers. She gave 2 flowers to Mary. How many flowers did she have left?

10 (−) 2 (=) 8 flowers

1 How many cats are there in all?

(A) 2

(B) 5

(C) 8

(D) 9

2 Tina picked these flowers.

How many flowers did Tina pick altogether?

(F) 35

(G) 8

(H) 7

(I) 5

3 Bob has 5 pennies. He gets 2 more. Which number sentence tells how many pennies he has in all?

(A) $5 - 2 = 3$ (C) $5 + 5 = 10$

(B) $5 + 2 = 7$ (D) $2 + 2 = 4$

4 What is the sum for 0 + 5?

F 0

G 5

H 6

I 50

MA.A.3.1.1.1.1

5 Carl has 2 toy trucks.

Kim has 3 toy trucks.

Who has more toy trucks?

Draw a picture to show how you found out.

_____

_ _ _ _ _ _ _ _ _ _ _ _ _ _ _

_____ has more toy trucks.

MA.A.3.1.2.1.3

1 Which matches the dots?

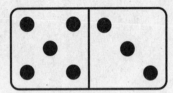

(A) 5 + 8 (C) 58

(B) 8 + 3 (D) 5 + 3

MA.A.3.1.1.1.1

2 Which is a way to make 10?

(F) 7 + 2 (H) 5 + 3

(G) 6 + 4 (I) 3 + 4

MA.A.3.1.1.1.2

3 Look at the first set of stars. Which picture shows the same number of stars?

(A) (B) (C) (D)

MA.A.3.1.1.1.4

4 Use cubes. Which two problems have the same sum?

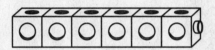

Ⓕ 1 + 0 and 2 + 1   Ⓗ 2 + 2 and 2 + 1

Ⓖ 1 + 0 and 2 + 2   Ⓘ 3 + 3 and 4 + 2

MA.A.3.1.2.1.2

5

THINK
SOLVE
EXPLAIN

Sam's Stamps	Carl's Stamps

How many stamps do Sam and Carl have in all?

Write a number sentence to solve.

_____ ◯ _____ ◯ _____ stamps

MA.A.3.1.2.1.4

1 Ben had 5 pennies. He lost 4 pennies.
How many pennies does he have left?

(A) 11

(C) 1

(B) 9

(D) 0

MA.A.3.1.1.1.1

2 Subtract. What is the difference?

(F) 8

(H) 4

(G) 6

(I) 2

MA.A.3.1.1.1.1

3 Jane had 6 books. She gave 1 to Ann.
How many books does Jane have left?

(A) 1

(C) 5

(B) 4

(D) 7

© Harcourt

MA.A.3.1.3.1.1

4 Mrs. Casson had 5 pencils.
She gave 2 pencils to Mike.
How many pencils are left?

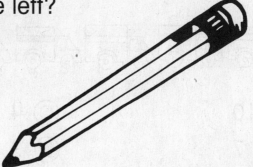

Ⓕ 7

Ⓖ 5

Ⓗ 3

Ⓘ 2

MA.A.3.1.3.1.1

5 6 people rode the bus.
3 people got off the bus.
How many people are left on the bus?

THINK
SOLVE
EXPLAIN

Draw a picture to solve.
Then write the number sentence.

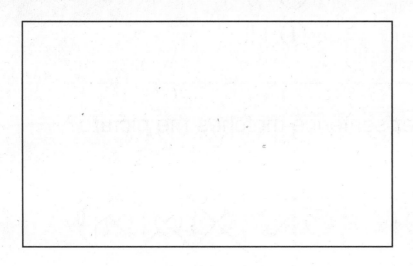

____ ◯ ____ ◯ ____ people

MA.A.3.1.2.1.4

© Harcourt

1 Ed has 7 toy trucks. He gives 3 to Beth.
How many toy trucks does Ed have left?

(A) 10   (C) 4

(B)   7   (D) 3

MA.A.3.1.1.1.1

2 How many more fish are there in the top row?

(F) 3   (H) 8

(G) 5   (I) 11

MA.A.3.1.1.1.1

3 Which number sentence matches the picture?

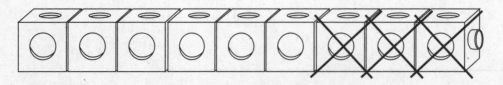

(A) 9 + 3 = 12   (C) 9 − 6 = 3

(B) 9 − 3 = 6   (D) 6 + 3 = 9

MA.A.3.1.2.1.4

4 Solve.

$8 - \blacksquare = 7$

(F) 1
(G) 2
(H) 3
(I) 4

MA.D.2.1.1.1.1

5 June had 6 stars. She gave some stars to Pete. She has 4 stars left. How many stars did she give away? Draw a picture to solve.

THINK
SOLVE
EXPLAIN

_____ stars

MA.A.3.1.1.1.1

1 Which does the number line show?

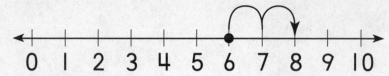

(A) 8 − 3

(B) 8 − 2

(C) 6 + 3

(D) 6 + 2

MA.A.3.1.1.1.1

2 Add.

$$\begin{array}{r} 3 \\ +\ 2 \\ \hline \end{array}$$

(F) 2

(G) 3

(H) 4

(I) 5

MA.A.3.1.1.1.2

3 Which number sentence shows this doubles fact?

(A) 1 + 2 = 3

(B) 2 + 2 = 4

(C) 4 + 3 = 7

(D) 5 + 5 = 10

MA.A.3.1.2.1.4

© Harcourt

4 Kathy has 7 cookies. Jill has 2 cookies.
How many cookies are there in all?

(F) 10   (H) 8

(G) 9   (I) 6

MA.A.3.1.3.1.1

5 Draw dots to show 9 in three
different ways.

THINK
SOLVE
EXPLAIN

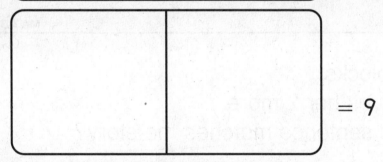
= 9

= 9

= 9

MA.A.3.1.2.1.2

© Harcourt

1 Which shows 5 + 4 another way?

● ● ● ● ●
○ ○ ○ ○

Ⓐ 4 + 3   Ⓒ 4 + 6

Ⓑ 4 + 5   Ⓓ 5 + 5

MA.A.3.1.1.1.4

2 Which shows the same fact as
$\begin{array}{r}4\\+3\\\hline\end{array}$?

Ⓕ $\begin{array}{r}3\\+3\\\hline\end{array}$ Ⓗ $\begin{array}{r}3\\+4\\\hline\end{array}$

Ⓖ $\begin{array}{r}4\\+2\\\hline\end{array}$ Ⓘ $\begin{array}{r}5\\+2\\\hline\end{array}$

MA.A.3.1.1.1.4

3 Hanna has 4 blocks.
Her teacher gives her 2 more.
Which number sentence matches the story?

Ⓐ 4 + 2 = 6   Ⓒ 4 + 4 = 8

Ⓑ 4 − 2 = 2   Ⓓ 6 − 4 = 2

MA.A.3.1.3.1.1

4 What is the missing number?

Add 3	
1	4
2	5
3	

(F) 1

(G) 2

(H) 3

(I) 6

MA.D.1.1.1.1.1

5 Write the rule.

 Explain your answer.

Add ___	
2	4
4	6
6	8

_ _ _ _ _ _ _ _ _ _ _ _ _ _ _ _

_____

_ _ _ _ _ _ _ _ _ _ _ _ _ _ _ _

_____

_ _ _ _ _ _ _ _ _ _ _ _ _ _ _ _

_____

© Harcourt

MA.D.1.1.1.1.2

1 Choose the related fact.

$$8 + 2 = 10$$

Ⓐ $10 - 2 = 8$ Ⓒ $10 - 7 = 3$

Ⓑ $10 - 9 = 1$ Ⓓ $10 - 3 = 7$

MA.A.3.1.1.1.1

---

2 Use the number line to solve.

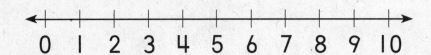

$$7 - 1 = \underline{\quad}$$

Ⓕ 2   Ⓗ 6

Ⓖ 5   Ⓘ 7

MA.A.3.1.1.1.2

---

3 Use the number line to solve.

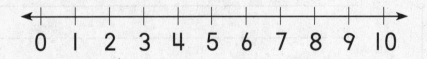

$$9 - 2 = \underline{\quad}$$

Ⓐ 6   Ⓒ 8

Ⓑ 7   Ⓓ 9

MA.A.3.1.1.1.2

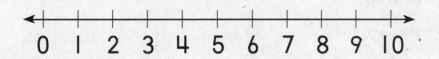
4 Use the number line to solve.

```
←—+—+—+—+—+—+—+—+—+—+—→
  0  1  2  3  4  5  6  7  8  9  10
```

8 – 3 = _____

Ⓕ 5
Ⓖ 4
Ⓗ 3
Ⓘ 2

MA.C.3.1.2.1.1

5 Draw a picture to solve.

 THINK SOLVE EXPLAIN

7 ducks were flying.
Some landed in a pond.
4 ducks are still in the air.
How many ducks landed?

_____ ducks

MA.A.3.1.3.1.1

1 Choose the subtraction sentence that tells about the picture.

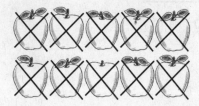

(A) $10 - 5 = 5$ (C) $10 - 0 = 10$

(B) $5 - 5 = 0$ (D) $10 - 10 = 0$

MA.A.3.1.1.1.1

2 Which number sentence finishes the fact family?

$$8 + 2 = 10$$
$$2 + 8 = 10$$
$$10 - 2 = 8$$

(F) $8 - 2 = 6$ (H) $10 - 8 = 2$

(G) $10 + 2 = 12$ (I) $10 + 8 = 18$

MA.A.3.1.1.1.3

3 Choose the rule.

Subtract ?	
7	5
8	6
9	7

(A) 2   (C) 4

(B) 3   (D) 5

MA.D.1.1.1.1.2

4 Choose the missing number.

$$\begin{array}{r} 3 \\ + \ \square \\ \hline 7 \end{array} \qquad \begin{array}{r} \square \\ + \ 3 \\ \hline 7 \end{array} \qquad \begin{array}{r} 7 \\ - \ \square \\ \hline 3 \end{array} \qquad \begin{array}{r} 7 \\ - \ 3 \\ \hline \square \end{array}$$

Ⓕ 3

Ⓖ 4

Ⓗ 7

Ⓘ 10

MA.D.2.1.1.1.1

5 Solve.

THINK
SOLVE
EXPLAIN

There are 9 crackers.

Children eat 5 of them.

How many are left?

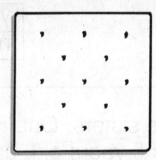

_____ crackers

Tell what you do to solve the problem.

_____

_____

_____

MA.A.3.1.2.1.1

1 Jake wants to know if his friends have pets. What question should he ask to find out?

Ⓐ Do you like pets?

Ⓑ Do you have pets?

Ⓒ Do you have a dog?

Ⓓ Is your cat nice?

MA.E.1.1.1.1.1

2 Fran asked three friends if they like milk. Which could show what she found out?

Ⓕ Like Milk?
yes

Ⓗ Like Milk?

Ⓖ Like Milk?
yes	I I
no	I

Ⓘ Like Milk?
yes	
no	

MA.E.1.1.1.1.2

3 Barb has three sisters. One is 2 years old, another is 5 years old, and the third is 10 years old. What is the difference in age between Barb's youngest sister and her oldest sister?

Ⓐ 3   Ⓒ 8

Ⓑ 7   Ⓓ 12

MA.E.1.1.2.1.1

© Harcourt

4 The class is planning what to eat at a picnic. Which survey question helps plan what to eat at a picnic?

Ⓕ What was the weather yesterday?

Ⓖ When do you want to have the picnic?

Ⓗ Where do you want to have the picnic?

Ⓘ What food do you want at the picnic?

MA.E.3.1.1.1.1

5 Amy recorded how much it rained each day last week. What is the best way for Amy to show the amount of rain for each day?

Show how you would share what Amy found out.

MA.E.3.1.2.1.2

1 Which tells how many there are?

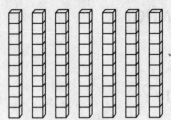

(A) 8 ones = 8   (C) 8 tens = 80

(B) 7 ones = 7   (D) 7 tens = 70

MA.A.2.1.1.1.2

2 Which number does the picture show?

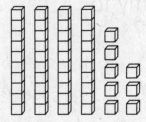

(F) 4   (H) 48

(G) 47   (I) 58

MA.A.2.1.2.1.1

3 Which is another way to group the blocks in the picture?

(A) 1 ten   (C) 8 tens

(B) 2 tens   (D) 10 tens

MA.A.2.1.2.1.2

Chapter 10 Practice Set

4 What is the value of the 8 in 83?

(F) 30

(G) 50

(H) 80

(I) 90

MA.A.2.1.2.1.3

5 Circle the closest estimate.

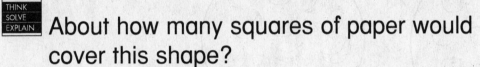

THINK
SOLVE
EXPLAIN

About how many squares of paper would cover this shape?

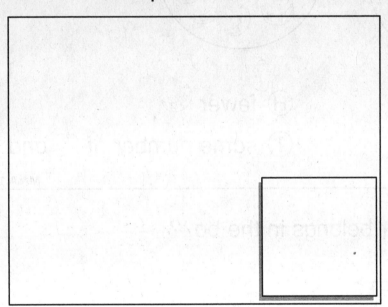

about 10   about 50   about 100

Draw a picture to show how you know.

© Harcourt

MA.A.4.1.1.1.3

1 Which number is between 85 and 87?

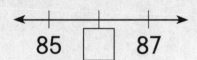

85 ☐ 87

(A) 83   (C) 86

(B) 84   (D) 88

MA.A.1.1.2.1.1

2 Which tells about these groups?

(F) more 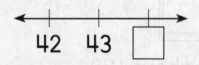   (H) fewer 🍌

(G) more 🍌   (I) same number of 🍎 and 🍌

MA.A.1.1.2.1.2

3 Which number belongs in the box?

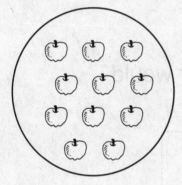

42   43   ☐

(A) 40   (C) 46

(B) 44   (D) 50

MA.C.3.1.2.1.1

4 Which symbol belongs in the circle?

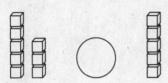

Ⓕ >

Ⓖ =

Ⓗ <

Ⓘ –

MA.D.2.1.1.1.2

5 Start on 56.

THINK SOLVE EXPLAIN

Count backward.

What nest does not have a bird in it?

Write the number.

Explain how you got your answer.

_____

- -

_____

- -

_____

MA.A.2.1.1.1.3

1 Count by fives. Which number is missing?

5, 10, 15, 20, 25, _____, 35

Ⓐ 50   Ⓒ 40

Ⓑ 45   Ⓓ 30

MA.A.2.1.1.1.1

2 Start at 6. Skip count by tens. Which shows the numbers you say?

1	2	3	4	5	6	7	8	9	10
11	12	13	14	15	16	17	18	19	20
21	22	23	24	25	26	27	28	29	30
31	32	33	34	35	36	37	38	39	40
41	42	43	44	45	46	47	48	49	50
51	52	53	54	55	56	57	58	59	60
61	62	63	64	65	66	67	68	69	70
71	72	73	74	75	76	77	78	79	80
81	82	83	84	85	86	87	88	89	90
91	92	93	94	95	96	97	98	99	100

Ⓕ 6, 16, 26, 36, 46, 56, 66, 76, 86, 96

Ⓖ 6, 7, 8, 9, 10, 11, 12, 13, 14, 15

Ⓗ 6, 10, 16, 20, 26, 30, 36, 40, 46, 50

Ⓘ 0, 10, 20, 30, 40, 50, 60, 70, 80, 90

MA.A.2.1.1.1.4

3 Which pattern does the hundred chart show?

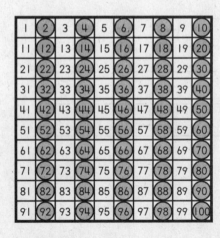

Ⓐ Count by twos.

Ⓑ Count by threes.

Ⓒ Count by fives.

Ⓓ Count by tens.

MA.D.1.1.1.1.3

© Harcourt

4 How many wheels are on 3 bikes?

number of bikes	1	2	3
number of wheels	2	4	?

Ⓕ 8

Ⓖ 7

Ⓗ 6

Ⓘ 5

MA.D.1.1.2.1.5

5 Circle the person that is sixth.
Explain your answer.

THINK
SOLVE
EXPLAIN

first

_____

- -

_____

- -

_____

MA.A.1.1.1.1.3

1 Draw lines to match.
What is the difference?

       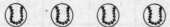

Ⓐ 0   Ⓒ 8

Ⓑ 1   Ⓓ 16

MA.A.3.1.1.1.1

2 What is the sum?

4 + 3 = _____

Ⓕ 1   Ⓗ 4

Ⓖ 3   Ⓘ 7

MA.A.3.1.1.1.2

3 What is the sum?

3 + 4 + 3 = _____

Ⓐ 12   Ⓒ 7

Ⓑ 10   Ⓓ 6

MA.A.3.1.1.1.4

© Harcourt

4 Which number sentence matches the picture?

Ⓕ 11 − 7 = 4
Ⓖ 10 − 4 = 6
Ⓗ 8 + 3 = 11
Ⓘ 5 + 6 = 11

MA.A.3.1.2.1.2

5 Draw a picture to solve.
Then write a number sentence.

THINK
SOLVE
EXPLAIN

7 cups of juice are on one table.
5 cups of juice are on another table.
How many cups of juice are there in all?

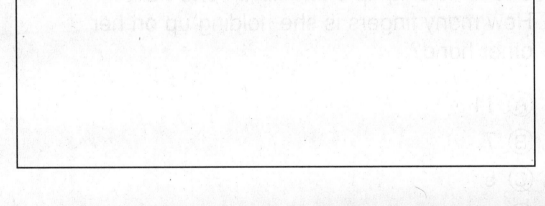

_____ ◯ _____ ◯ _____ cups of juice

MA.A.3.1.2.1.4

© Harcourt

1 What is the fact family for 6, 3, and 3?

Ⓐ $3 - 3 = 0$ and $3 + 3 = 6$

Ⓑ $3 + 3 = 6$ and $6 + 3 = 9$

Ⓒ $3 + 3 = 3$ and $3 + 3 + 6 = 12$

Ⓓ $3 + 3 = 6$ and $6 - 3 = 3$

MA.A.3.1.1.1.3

2 Doug has 12 hats. He gives 5 away.
How many hats are left?

Ⓕ 12

Ⓖ 10

Ⓗ 8

Ⓘ 7

MA.A.3.1.2.1.1

3 Pat is holding up 8 fingers.
She is holding up 3 fingers on one hand.
How many fingers is she holding up on her
other hand?

Ⓐ 11

Ⓑ 7

Ⓒ 5

Ⓓ 3

MA.A.3.1.3.1.1

© Harcourt

Name _____

4 What is the missing number?

6 − [ ] = 2

Ⓕ 8

Ⓖ 6

Ⓗ 4

Ⓘ 2

MA.D.2.1.1.1.1

5 Choose a way to solve the problem.
Show your work.

THINK
SOLVE
EXPLAIN

7 ducks are in the water.
3 ducks join them.
How many ducks are there now?

_____ ducks

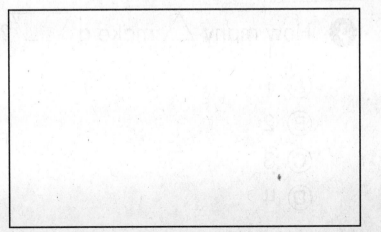

MA.A.3.1.2.1.3

© Harcourt

Name_____

1 How many sides and vertices does a square have?

   Ⓐ 5 sides and 5 vertices

   Ⓑ 4 sides and 4 vertices

   Ⓒ 3 sides and 3 vertices

   Ⓓ 2 sides and 2 vertices

MA.C.1.1.1.1.1

2 Which shows a face for this solid?

  Ⓕ   Ⓗ

  Ⓖ   Ⓘ

MA.C.1.1.1.1.2

3 How many 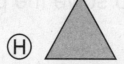 make a  ?

   Ⓐ 1

   Ⓑ 2

   Ⓒ 3

   Ⓓ 4

MA.C.2.1.1.1.2

4 What shape is the door?

Ⓕ rectangle

Ⓖ square

Ⓗ triangle

Ⓘ circle

MA.C.3.1.1.1.2

5 Color the objects that have 2 faces and
are curved.
Tell how you know.

THINK
SOLVE
EXPLAIN

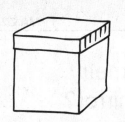  

_____

_ _

_____

_ _

_____

© Harcourt

MA.C.3.1.1.1.3

1 In which set are all the figures triangles?

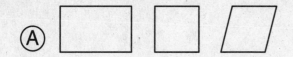

Ⓐ

Ⓑ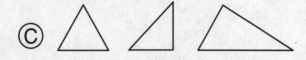

Ⓒ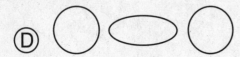

Ⓓ

MA.C.1.1.1.1.3

2 Which shape shows a line of symmetry?

Ⓕ 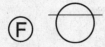 Ⓗ

Ⓖ  Ⓘ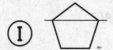

MA.C.2.1.1.1.1

3 Karen holds a mirror on the line of symmetry. Which shows what she will see in the mirror?

Ⓐ   Ⓒ

Ⓑ   Ⓓ

MA.C.2.1.1.1.3

© Harcourt

4 Which picture shows a slide?

Ⓕ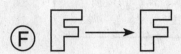

Ⓗ

Ⓖ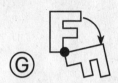

Ⓘ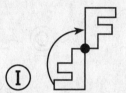

MA.C.2.1.2.1.1

5

 THINK SOLVE EXPLAIN

Draw a bird to the right of the tree.

Draw a sun over the tree.

Draw a cloud to the left of the tree.

Draw a dog beside the tree.

MA.C.2.1.1.1.4

1 Which shows the right way to continue the pattern?

Ⓐ  Ⓒ

Ⓑ  Ⓓ

MA.D.1.1.1.1.1

2 What comes next in the pattern?

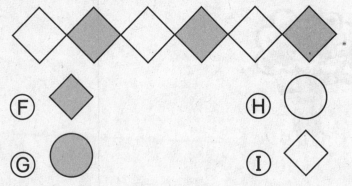

Ⓕ Ⓗ

Ⓖ Ⓘ

MA.D.1.1.1.1.4

3 Val made a pattern using these shapes.
Which shows the pattern that Val made?

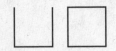

Ⓐ

Ⓑ

Ⓒ

Ⓓ

MA.D.1.1.2.1.1

© Harcourt

4 Which is another way to show this pattern?

F   I, I, I, I, I, I, I, I, I

G   I, I, 2, I, I, 2, I, I, 2

H   I, 2, 2, I, 2, 2, I, 2, 2

I   2, 2, 2, 2, 2, 2, 2, 2, 2

MA.D.1.1.2.1.2

5 Find the pattern.

Draw what comes next.

Explain what happens in the pattern.

_____

– – – – – – – – – – – – – – – – – –

_____

_____

– – – – – – – – – – – – – – – – – –

_____

MA.D.1.1.2.1.3

Chapter 17 Practice Set

1 Which model shows 11 counters?

Ⓐ

Ⓒ

Ⓑ

Ⓓ

MA.A.1.1.3.1.1

2 The basketball team has 6 girls and 5 boys.
How many people are on the team?

Ⓕ 8

Ⓗ 11

Ⓖ 10

Ⓘ 12

MA.A.3.1.1.1.2

3 Peter has 10 pennies and 5 nickels.
How many coins does Peter have?

Ⓐ 5

Ⓒ 15

Ⓑ 14

Ⓓ 35

MA.A.3.1.3.1.1

4 What number can you write in the box
to make this number sentence true?

$7 + \boxed{} = 12$

Ⓕ 5
Ⓖ 6
Ⓗ 9
Ⓘ 10

MA.D.2.1.1.1.1

5 Add these numbers in two ways.
Circle the numbers you add first.
Write the sums.
Tell how you solved each problem.

| THINK |
| SOLVE |
| EXPLAIN |

$\begin{array}{r} 7 \\ 3 \\ +4 \\ \hline \end{array}$ $\begin{array}{r} 7 \\ 3 \\ +4 \\ \hline \end{array}$

_____

_____

_____

_____

_____

MA.A.3.1.1.1.4

© Harcourt

1 Danny has 8 stickers. He got 8 more stickers.
How many stickers does Danny have now?

Ⓐ 0

Ⓑ 8

Ⓒ 15

Ⓓ 16

MA.A.3.1.1.1.2

2 Which is the sum?

$$8$$
$$+3$$

Ⓕ 10

Ⓖ 11

Ⓗ 12

Ⓘ 13

MA.A.3.1.1.1.2

3 Shelly has 12 pens. She gives 6 pens to Tim.
How many pens does Shelly have left?

Ⓐ 6

Ⓑ 8

Ⓒ 9

Ⓓ 10

MA.A.3.1.1.1.2

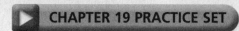

Choose the best estimate.

4 Charlie has 18 cards.
He gives 9 cards to Jim.
About how many cards does
Charlie have left?

Ⓕ Charlie has more than 20 cards.

Ⓖ about 2

Ⓗ about 10

Ⓘ about 20

MA.A.4.1.1.1.4

5 Write the number sentence that tells about
the number line.

THINK
SOLVE
EXPLAIN

Explain your answer.

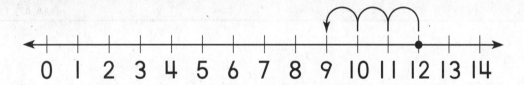

____ ◯ ____ ◯ ____

_____

_ _ _ _ _ _ _ _ _ _ _ _ _ _ _ _ _ _ _

_____

_ _ _ _ _ _ _ _ _ _ _ _ _ _ _ _ _ _ _

_____

MA.A.3.1.1.1.2

1 Which picture shows the number 17?

Ⓐ

Ⓒ

Ⓑ

Ⓓ

MA.A.1.1.4.1.1

2 Which numbers are in this fact family?

$$5 + 7 = 12 \qquad 7 + 5 = 12$$
$$12 - 5 = 7 \qquad 12 - 7 = 5$$

Ⓕ 2, 5, 7

Ⓗ 1, 2, 5, 7

Ⓖ 1, 2, 3

Ⓘ 5, 7, 12

MA.A.3.1.1.1.3

3 Use ⬜.
Which picture shows this problem?

$$\begin{array}{r} 5 \\ + 6 \\ \hline \end{array}$$

Ⓐ

Ⓑ

Ⓒ

Ⓓ

MA.A.3.1.2.1.2

© Harcourt

4 There are 20 apples. 12 apples are big. The rest are small. How many apples are small?

(F) 32

(G) 12

(H) 8

(I) 6

MA.A.3.1.3.1.1

5 THINK SOLVE EXPLAIN There are 19 children in Ms. Chin's class. 18 are here today. How many children are absent? Show how you can solve the problem.

MA.D.2.1.2.1.1

1 What fraction of the pizza is left?

Ⓐ $\frac{1}{1}$ Ⓒ $\frac{1}{3}$

Ⓑ $\frac{1}{2}$ Ⓓ $\frac{1}{4}$

MA.A.1.1.3.1.2

2 Which shows $\frac{1}{2}$ of the bananas shaded?

Ⓕ 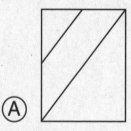   Ⓗ

Ⓖ 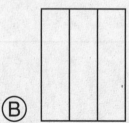   Ⓘ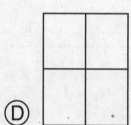

MA.A.1.1.3.1.2

3 Ted folded his paper in thirds.
Which shows thirds?

Ⓐ   Ⓒ

Ⓑ   Ⓓ

MA.A.1.1.3.1.3

© Harcourt

4 Jerry wants to cut a cookie to show halves.
How many halves make one whole?

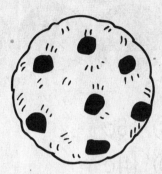

Ⓕ 1   Ⓗ 3
Ⓖ 2   Ⓘ 4

MA.A.1.1.3.1.4

5 Read the clues. Draw lines in the pictures to
show how the children cut the pizzas.

THINK
SOLVE
EXPLAIN

Cara cuts a pizza.   Ted cuts a pizza.
It has 4 equal slices.   It has 2 equal slices.

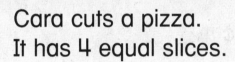   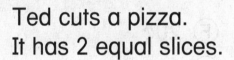

   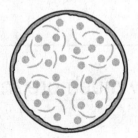

Who has the pizza with more slices, Cara or Ted?

_____

Who has the pizza with bigger slices, Cara or Ted?

_____

MA.A.1.1.3.1.3

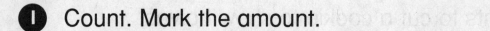

1 Count. Mark the amount.

Ⓐ 35¢   Ⓒ 15¢

Ⓑ 21¢   Ⓓ 14¢

MA.B.3.1.1.1.3

2 Count. Mark the amount.

Ⓕ 20¢   Ⓗ 40¢

Ⓖ 30¢   Ⓘ 50¢

MA.B.3.1.1.1.3

3 How much are these coins worth?

Ⓐ 52¢   Ⓒ 27¢

Ⓑ 37¢   Ⓓ 5¢

MA.B.3.1.1.1.3

© Harcourt

4 Tara wants to buy a book for 16¢.
Which group of coins could she use?

Ⓕ

Ⓖ

Ⓗ

Ⓘ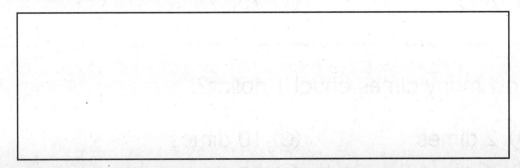

MA.A.1.1.4.1.1

5 These are Beth's coins.

 THINK SOLVE EXPLAIN

Tim has the same amount of money in nickels.
Draw the nickels Tim has.

How much are Tim's nickels worth? _____ ¢

MA.B.3.1.1.3

© Harcourt

1 Which group of coins shows the same amount as the pennies?

Ⓐ

Ⓒ

Ⓑ

Ⓓ

MA.B.3.1.1.1.3

2 How much are these coins worth?

Ⓕ 40¢   Ⓗ 70¢

Ⓖ 65¢   Ⓘ 75¢

MA.B.3.1.1.1.3

3 How many dimes equal 1 dollar?

Ⓐ 2 dimes   Ⓒ 10 dimes

Ⓑ 5 dimes   Ⓓ 25 dimes

MA.B.3.1.1.1.3

4 Which is another way to show 35¢?

F

G

H

I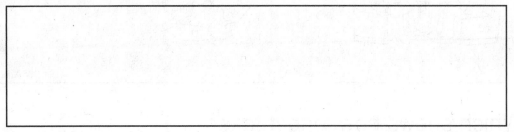

MA.B.3.1.1.1.3

5 Val has I quarter, I penny, I nickel, I half dollar, and I dime. Draw Val's coins in order from least value to greatest value.

THINK
SOLVE
EXPLAIN

least value   greatest value

How much are Val's coins worth?

_____¢

© Harcourt

MA.B.3.1.1.1.3

Name

1 The clock shows when the movie will start.

At what time will the movie start?

(A) 6:00　　(C) 7:30

(B) 6:30　　(D) 8:30

MA.B.1.1.1.1.4

2 Which takes about half an hour?

(F) paint a house

(H) clap your hands

(G) eat dinner

(I) write your name

MA.B.3.1.1.1.2

3 Which shows how long it takes to write the numbers 1 to 10?

(A) 1 week　　(C) 1 hour

(B) 1 day　　(D) 1 minute

MA.B.4.1.1.1.1

© Harcourt

4 Tomi wants to know what time it is.
Which tool should he use?

(F)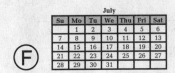

(G)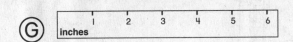

(H)

(I)

MA.B.4.1.2.1.2

5 Draw the hour hand and the minute hand on
the clock.

THINK
SOLVE
EXPLAIN

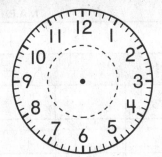   

What time will it be one hour after
the time on the clock?

_____:_____

MA.B.1.1.1.1.4

© Harcourt

Chapter 24 Practice Set

1 Which tool can you use to measure days?

Ⓐ

Ⓒ

Ⓑ

Ⓓ

MA.B.4.1.2.1.2

2 Which activity lasts the longest amount of time?

Ⓕ biking   Ⓗ eating

Ⓖ singing   Ⓘ swimming

Activity	Start	End
biking		
singing		

MA.E.1.1.1.1.4

3

Our Favorite Fall Month						
September	🍂	🍂	🍂	🍂		
October	🎃	🎃				
November	🦃	🦃	🦃	🦃	🦃	

0   1   2   3   4   5   6

Which month did the most children choose?

Ⓐ September   Ⓒ November

Ⓑ October   Ⓓ December

MA.E.3.1.1.1.2

© Harcourt

4

Our Favorite Time		Total
morning	ЖЖ II	7
afternoon	ЖЖ I	6
evening	ЖЖ ЖЖ	10

How many children chose afternoon or evening?

(F) 7 (H) 16

(G) 13 (I) 17

MA.E.3.1.1.1.3

5 Finish the tally table. Make a picture graph.

THINK
SOLVE
EXPLAIN

Our Favorite Summer Month		Total
June	ЖЖ	
July	ЖЖ IIII	
August	ЖЖ II	

Our Favorite Summer Month										
June										
July										
August										

0 1 2 3 4 5 6 7 8 9 10

Which month did the fewest children choose?

_____

MA.E.1.1.1.1.3

© Harcourt

1 Which pencil is the longest?

Ⓐ

Ⓑ

Ⓒ

Ⓓ

MA.B.1.1.1.1.1

2 Which estimate is the closest measurement for how tall your classroom door might be?

Ⓕ about 1 inch   Ⓗ about 1 foot

Ⓖ about 7 inches   Ⓘ about 7 feet

MA.B.1.1.1.1.2

3 Which toy is the shortest?

Ⓐ

Ⓑ

Ⓒ

Ⓓ

MA.B.2.1.1.1.2

© Harcourt

4 The marker is 9 centimeters long.
Which is the best estimate for the length of
the crayon?

Ⓕ about 3 centimeters

Ⓖ about 7 centimeters

Ⓗ about 9 centimeters

Ⓘ about 11 centimeters

MA.B.3.1.1.1.1

5 About how many beads long is the string?

THINK
SOLVE
EXPLAIN

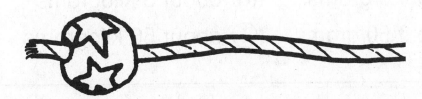

about _____ beads

Draw the beads on the string to show your answer.

MA.A.4.1.1.1.2

1 Which would you measure in kilograms?

(A)

(B)

(C)

(D)

MA.B.1.1.1.1.3

2 A paper clip weighs about 1 gram.

Which is the best estimate for how much a nickel weighs?

(F) about 5 grams (H) about 5 kilograms

(G) about 50 grams (I) about 50 kilograms

MA.B.1.1.2.1.1

3 Which weighs more than 1 pound?

(A)

(B)

(C)

(D)

MA.B.2.1.1.1.3

© Harcourt

4 This butter weighs about 1 pound.

Which is the best estimate for the weight of this bag of potatoes?

Ⓕ less than 1 pound

Ⓖ about 1 pound

Ⓗ about 10 pounds

Ⓘ more than 100 pounds

MA.B.2.1.2.1.1

5 Vera weighed these bags of apples.

 THINK SOLVE EXPLAIN

Draw the bags of apples in order from the lightest to the heaviest.

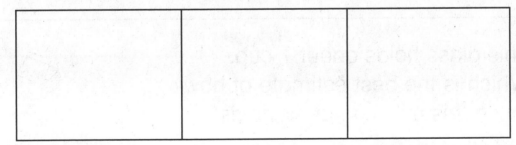

| lightest | | heaviest |

Explain how you know which bag is lightest.

_____

_____

MA.B.2.1.1.1.1

© Harcourt

1 What is the temperature?

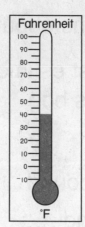

Fahrenheit

Ⓐ 50°F

Ⓑ 40°F

Ⓒ 30°F

Ⓓ 0°F

MA.B.1.1.1.1.5

2 Which container would you use to measure a quart of juice?

Ⓕ

Ⓗ

Ⓖ

Ⓘ

MA.B.1.1.1.1.6

3 This glass holds about 1 cup. Which is the best estimate of how much this milk container holds?

Ⓐ less than 1 cup

Ⓑ about 1 cup

Ⓒ about 4 cups

Ⓓ more than 10 cups

MA.B.2.1.2.1.1

© Harcourt

4 Which could you use to measure how much this mug holds?

 ⒡ 　　ⓗ

ⓖ 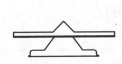　　ⓘ

MA.B.4.1.1.1.1

5 Kevin wants to measure the length of this pencil.

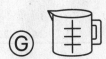
THINK
SOLVE
EXPLAIN
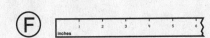

Circle the tool he should use.

  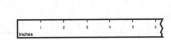

Draw to show how you would use the tool.

MA.B.4.1.2.1.1

1 What is the sum of 29 + 1?

tens	ones
2	9
+	1

tens	ones

- (A) 12
- (B) 29
- (C) 30
- (D) 31

MA.A.2.1.2.1.2

2 Ron adds two numbers.

$$\begin{array}{r} 50 \\ +30 \\ \hline 80 \end{array}$$

What does 80 mean?

- (F) 8 ones
- (G) 18 ones
- (H) 8 tens
- (I) 18 tens

MA.A.2.1.2.1.3

3 Lori buys two toys.

$$\begin{array}{r} 54¢ \\ + 24¢ \\ \hline ¢ \end{array}$$

How much do the toys cost in all?

- (A) 74¢
- (B) 78¢
- (C) 84¢
- (D) 88¢

MA.A.3.1.1.1.5

© Harcourt

4 Which is the best estimate?
Paul counts 22 red birds. He counts 8 black birds. Paul counts 29 blue birds.
About how many birds does he count?

Ⓕ about 6 birds   Ⓗ about 60 birds

Ⓖ about 20 birds   Ⓘ about 200 birds

MA.A.4.1.1.1.1

5 Use Workmat 3 and 🎲 to subtract.

THINK
SOLVE
EXPLAIN

Mike has 27 baseball cards.
He gives 4 cards to Jason.
How many cards does Mike have left?

Draw 🎲 to show your answer.

tens	ones
2	7
−	4

┌─Workmat─┐

Tens	Ones

Mike has _____ baseball cards left.

MA.A.1.1.3.1.1

1 Sal spins the pointer 10 times. He makes these tally marks.

Tally Marks		Total
	ҢНI	
	III	
●	II	

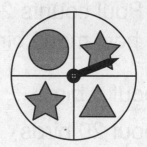

Predict. If he spins the pointer 10 more times, on which shape will it likely stop most often?

Ⓐ ★

Ⓑ ▲

Ⓒ ●

Ⓓ ■

MA.E.1.1.3.1.1

2 Which completes the sentence?

I am most likely to pull a _____.

Ⓕ ●

Ⓖ 🌀

Ⓗ ○

Ⓘ ●

MA.E.2.1.1.1.1

3 Which completes the sentence?

It is impossible for the pointer to stop on _____.

Ⓐ 1

Ⓑ 2

Ⓒ 3

Ⓓ 4

MA.E.2.1.1.1.2

Chapter 30 Practice Set

4 Which buttons are you equally likely to pull?

  (F)  (H)

  (G)   (I)

MA.E.2.1.2.1.1

5 Color the spinner to the match this clue.

It is less likely that the pointer will stop on red than on blue.

How do you know your spinner is correct?

_____

_ _

_____

_ _

_____

MA.E.2.1.1.1.3

1 Which addition sentence matches the story?

There are 4 turtles.
2 turtles join them.
How many turtles are there in all?

Ⓐ 6 − 2 = 4   Ⓒ 4 + 2 = 6
Ⓑ 6 − 4 = 2   Ⓓ 6 + 2 = 8

MA.A.3.1.2.1.4

2 How much money is shown?

Ⓕ 41¢   Ⓗ 51¢
Ⓖ 43¢   Ⓘ 52¢

MA.B.3.1.1.1.3

3 Which number has a 2 in the tens place?

Ⓐ 2   Ⓒ 29
Ⓑ 12   Ⓓ 200

MA.A.2.1.2.1.3

Go On ▶

© Harcourt

Name_____

4 Which shows the right way to continue the pattern?

Ⓕ △ ▼

Ⓖ ▼ ▼

Ⓗ ▼ △

Ⓘ △ ▽

MA.D.1.1.1.1.1

5 Paul asked 6 friends if they have sisters.
3 friends said yes. 3 friends said no.
Which graph shows what Paul found out?

Ⓐ
Sisters

| yes | 🧍 | 🧍 | 🧍 |
| no | 🧍 | 🧍 | 🧍 |

Ⓒ
Sisters

| yes | 🧍 | 🧍 | 🧍 |
| no | 🧍 | | |

Ⓑ
Sisters

| yes | 🧍 | 🧍 | 🧍 |
| no | 🧍 | 🧍 | |

Ⓓ
Sisters

| yes | 🧍 | 🧍 |
| no | 🧍 | 🧍 |

MA.E.1.1.1.1.3

6 Count back. Which is the difference?

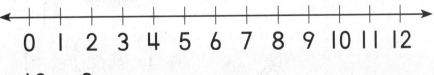

12 − 3 = _____

Ⓕ 9   Ⓗ 7

Ⓖ 8   Ⓘ 1

MA.A.3.1.1.1.2

Go On ▶

© Harcourt

7 Kim wants to know how many days are in 1 month. Which tool should she use?

Ⓐ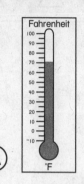

Ⓒ

Ⓑ

Ⓓ

MA.B.4.1.2.1.2

8 Which number belongs in the box?

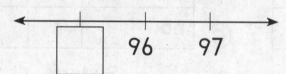

Ⓕ 91   Ⓗ 95

Ⓖ 94   Ⓘ 99

MA.C.3.1.2.1.1

9 Which toy is impossible for Karen to pick?

Ⓐ   Ⓒ

Ⓑ   Ⓓ

MA.E.2.1.1.1.2   **Go On ▶**

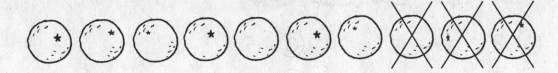

Name_____

10 Vince bought 10 oranges.
He gave 3 oranges to his friend.
How many oranges are left?

(F) 3   (H) 10

(G) 7   (I) 13

MA.A.3.1.1.1.1

11 Carol estimated the length of her bicycle.
Which is a good estimate?

(A) 2 inches   (C) 5 inches

(B) 2 feet   (D) 5 feet

MA.B.3.1.1.1.1

12 A bowl has 3 green cubes and 9 red cubes. Todd predicts that he will pull a green cube. Which is true?

(F) This is less likely.   (H) This is certain.

(G) This is more likely.   (I) This is impossible.

MA.E.2.1.1.1.1   **Go On** ▶

© Harcourt

13 How many ladybugs are there in all?

Ⓐ 3   Ⓒ 9

Ⓑ 7   Ⓓ 10

MA.A.1.1.1.1.1

14 Sam sorted objects. Some objects are round. Some objects are not round. Which shows all round objects?

Ⓕ

Ⓖ

Ⓗ

Ⓘ

MA.C.3.1.1.1.1

15 Nancy puts 50 pictures in her book. She puts 5 pictures on each page. How many pages does she fill?

Ⓐ 55   Ⓒ 20

Ⓑ 45   Ⓓ 10

MA.A.1.1.3.1.1   **Go On ▶**

16 Felix drew this pattern.

Which rule did Felix use to draw the pattern?

MA.D.1.1.1.1.2

17 How long is the pencil?

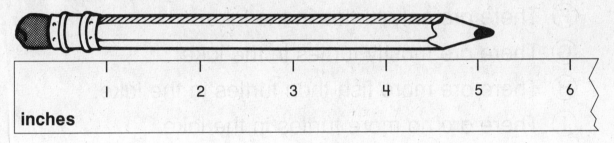

Ⓐ about 3 inches   Ⓒ about 5 inches

Ⓑ about 4 inches   Ⓓ about 6 inches

MA.B.1.1.2.1.1

18 Which is NOT a way to show 13?

Ⓕ

Ⓖ [image of base-ten blocks]

Ⓗ thirty

Ⓘ 1 ten and 3 ones

MA.A.1.1.4.1.1   **Go On ▶**

© Harcourt

19 Which is the missing number?

$$
\begin{array}{r}
21 \\
+\ \square \\
\hline
25
\end{array}
$$

Ⓐ 45

Ⓒ 6

Ⓑ 23

Ⓓ 4

MA.D.2.1.1.1.1

20 Beth went fishing with her dad. She caught 1 turtle and 6 fish. Which do you think is true?

Ⓕ There are only turtles in the lake.

Ⓖ There are mostly turtles in the lake.

Ⓗ There are more fish than turtles in the lake.

Ⓘ There are no more turtles in the lake.

MA.E.1.1.3.1.1

21 Which picture shows a turn?

MA.C.2.1.2.1.1

Go On ▶

© Harcourt

22 Which shows one whole shaded?

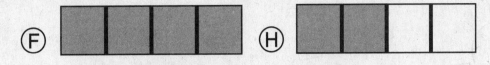

Ⓕ ⬜⬜⬜⬜   Ⓗ ⬜⬜⬜⬜

Ⓖ ⬜⬜⬜⬜   Ⓘ ⬜⬜⬜⬜

MA.A.1.1.3.1.4

23 Doug wants to measure how much water his glass holds. Which should he use?

Ⓐ cups

Ⓒ gallons

Ⓑ quarts

Ⓓ liters

MA.B.1.1.1.1.6

24 Without adding or subtracting, circle the best estimate.

Patti had 40 dolls.
She gave about 8 to Julie.
About how many does she have left?

Ⓕ about 10

Ⓖ about 20

Ⓗ about 30

Ⓘ about 40

MA.A.4.1.1.1.1

© Harcourt

STOP

1 How much do you spend for both?

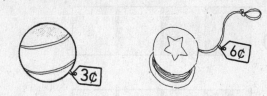

Ⓐ 3¢   Ⓒ 9¢

Ⓑ 6¢   Ⓓ 10¢

MA.A.3.1.1.1.1

2 Kathy counted the dogs and cats she saw.
Which question can you answer about her graph?

Animals I Saw		Total
cats	ЖЖ II	7
dogs	ЖЖ	5

Ⓕ How many fish did she see?

Ⓖ How many people did she see?

Ⓗ How many more cats did she see than dogs?

Ⓘ When did she see the cats and dogs?

MA.E.3.1.2.1.2

3 Which doubles fact matches the picture?

Ⓐ 5 + 5 = 10   Ⓒ 3 + 3 = 6

Ⓑ 4 + 4 = 8   Ⓓ 2 + 2 = 4

MA.A.3.1.1.1.2   **Go On ▶**

4 How are the shapes sorted?

(F) little and big

(G) curved and not curved

(H) shaded and not shaded

(I) triangles and rectangles

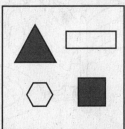

MA.C.3.1.1.1.3

---

5 How many hands are on 5 people?

number of people	1	2	3	4	5
number of hands	2	4	6	8	?

(A) 5   (C) 9

(B) 6   (D) 10

MA.D.1.1.2.1.5

---

6 Which number sentence solves the problem?

Nan has 10 pens. 3 pens are red.

How many are **not** red?

(F) $10 + 3 = 13$ (H) $10 - 10 = 0$

(G) $10 - 3 = 7$ (I) $3 + 3 = 6$

MA.A.3.1.2.1.4   **Go On ▶**

7 Tim needs to go to school tomorrow.
Which tells when he should go to bed?

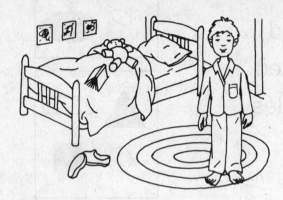

Ⓐ morning   Ⓒ afternoon

Ⓑ noon   Ⓓ evening

MA.B.3.1.1.1.2

8 Which is another way to write 2 tens
and 7 ones?

Ⓕ 271   Ⓗ 27

Ⓖ 207   Ⓘ 17

MA.A.1.1.1.1.2

9 Frank has two brothers that are 3 years old.
He has another brother that is 9 years old.
How old are most of his brothers?

Ⓐ 2   Ⓒ 9

Ⓑ 3   Ⓓ 12

MA.E.1.1.2.1.1   **Go On** ▶

© Harcourt

Name_____

10 Which does the picture show?

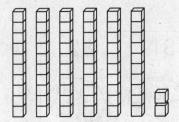

(F) 8 ones

(G) 6 tens and 0 ones

(H) 6 tens and 2 ones

(I) 8 tens and 2 ones

MA.A.2.1.2.1.1

11 Which has 3 vertices and 3 sides?

(A) ▭   (C) ◯

(B) △   (D) ▢

MA.C.1.1.1.1.1

12 What comes next in this pattern?

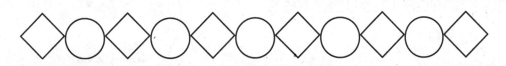

(F) ◯   (H) ◆

(G) ◇   (I) ●

MA.D.1.1.1.1.4   **Go On** ▶

13 Count backward.
Which number is the black car?

80   81   82

Ⓐ 82   Ⓒ 78

Ⓑ 79   Ⓓ 2

MA.A.2.1.1.1.3

14 Which is the longest object?

Ⓕ

Ⓖ ERASER

Ⓗ

Ⓘ

MA.B.2.1.1.1.1

15 Suzie spins the pointer.

Which shape is Suzie less
likely to spin?

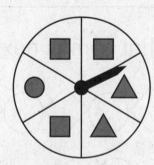

Ⓐ   Ⓒ

Ⓑ   Ⓓ

© Harcourt

16 Which frog is jumping?

first

Ⓕ second   Ⓗ fourth

Ⓖ third   Ⓘ seventh

MA.A.1.1.1.1.3

17 What is the temperature?

Ⓐ 50°F

Ⓑ 40°F

Ⓒ 30°F

Ⓓ 0°F

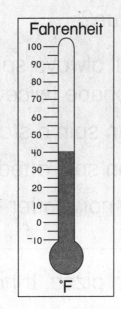

Fahrenheit

MA.B.1.1.1.1.5

18 There are 18 players on the team.
10 are girls. How many are boys?

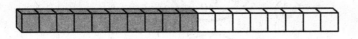

Ⓕ 8   Ⓗ 18

Ⓖ 10   Ⓘ 28

MA.D.2.1.2.1.1   **Go On** ▶

19 Roger has 9 toy cars.
He gets 9 more toy trucks.
About how many toy trucks
does he have now?

Ⓐ about 5   Ⓒ about 15

Ⓑ about 10   Ⓓ about 20

MA.A.4.1.1.1.4

20 Look at the spinner.
Which is true?

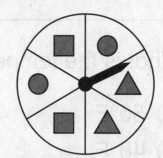

Ⓕ You will always spin the
same shape twice in a row.

Ⓖ You can spin a star.

Ⓗ You can spin a teddy bear.

Ⓘ It does not matter who spins the pointer.

MA.E.2.1.1.1.3

21 Pam cuts a pizza. It has 2 parts.
The parts are not equal. Which is Pam's
pizza?

Ⓐ

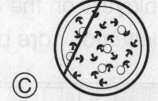

Ⓒ

Ⓑ

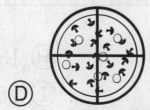

Ⓓ

MA.A.1.1.3.1.3

Go On ▶

22 Ellen wants to know the temperature. Which tool should she use?

(F) 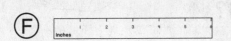 inches

(H)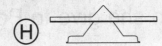

(G)
Fahrenheit

(I)

MA.B.4.1.2.1.1

23 Which object is to the left of the dish?

(A)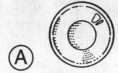

(C)

(B)

(D)

MA.C.2.1.1.1.4

24 Which is the difference?

tens	ones
9	5
−	4

(F) 7

(H) 61

(G) 60

(I) 91

STOP

MA.A.3.1.1.1.5

1 Kim had 10 pieces of candy in a jar.
She gave some to Pat.
Kim has 7 pieces of candy left.
How many did she give to Pat?

(A) 107   (C) 4

(B) 17   (D) 3

MA.A.3.1.1.1.1

2 Look at the graph.

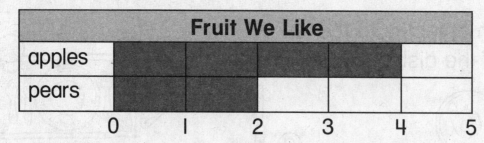

Fruit We Like					
apples					
pears					

0   1   2   3   4   5

How many fewer pears than apples did the children like?

(F) 2   (H) 6

(G) 4   (I) 10

MA.E.1.1.1.1.4

3 Which is the missing number?

$8 - \boxed{} = 3$

(A) 103   (C) 8

(B) 13   (D) 5

MA.D.2.1.1.1.1   **Go On ▶**

© Harcourt

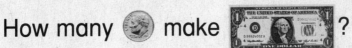
4 How many make ?

(F) 5

(G) 8

(H) 9

(I) 10

MA.B.3.1.1.1.3

5

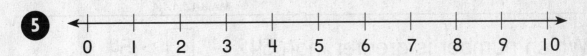

There are 8 pencils in a box. 2 fall out.
How many pencils are in the box now?

(A) 11   (C) 8

(B) 10   (D) 6

MA.A.3.1.1.1.2

6 Which shape is the window?

(F) rectangle   (H) triangle

(G) square   (I) circle

MA.C.3.1.1.1.2   **Go On ▶**

© Harcourt

7 Kim set her clock for 7:30.
Which shows her clock?

Ⓐ

Ⓑ

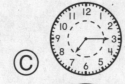

Ⓒ

Ⓓ

MA.B.1.1.1.1.4

8 Which number is greater than 54?   ☐ > 54

Ⓕ 68   Ⓗ 49

Ⓖ 53   Ⓘ 30

MA.A.1.1.2.1.2

9 5 friends told Jay they like red.
2 friends like green.
1 friend likes blue.
Which shows what Jay found out?

Ⓐ

Colors We Like		Total
red	|||	3
green	||	2
yellow	⊬⊬⊬	5

Ⓒ

Colors We Like		Total
red	⊬⊬⊬	5
green	||	2
blue			1

Ⓑ

Colors We Like		Total
red	⊬⊬⊬	5
green			1
blue	||	2

Ⓓ

Colors We Like		Total
yellow	||	2
blue	⊬⊬⊬	5
red			1

MA.E.3.1.1.1.2

Go On ▶

Practice Test 3

Name_____

10 Bill has 50 toy cars. Sam has 10 less. How many toy cars does Sam have?

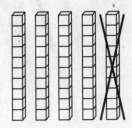

(F) 10   (H) 30

(G) 20   (I) 40

MA.A.3.1.2.1.2

11 How many vertices and sides does a triangle have?

(A) 6 vertices and 6 sides

(B) 5 vertices and 5 sides

(C) 4 vertices and 4 sides

(D) 3 vertices and 3 sides

MA.C.1.1.1.1.3

12 Ben made this pattern.

Which continues Ben's pattern?

(F)    (H) ♡ ♡

(G) 🖤 🖤   (I) ♡ ♡

MA.D.1.1.2.1.3   **Go On ▶**

13 Soccer practice starts at 3:00.
What time of the day is soccer practice?

Ⓐ morning

Ⓑ afternoon

Ⓒ evening

Ⓓ night

MA.B.3.1.1.1.2

14 Which model shows an odd number?

Ⓕ

Ⓖ

Ⓗ

Ⓘ

MA.A.5.1.1.1.1

15 How many △ make a ⬡ ?

Ⓐ 3

Ⓑ 6

Ⓒ 8

Ⓓ 10

MA.C.2.1.1.1.2   **Go On ▶**

Name _____

16 Which is the sum?

9
3
+ 6

Ⓕ 20   Ⓗ 16
Ⓖ 18   Ⓘ 15

MA.A.3.1.1.1.4

17 Which pencil is the shortest?

Ⓐ

Ⓑ

Ⓒ

Ⓓ

MA.B.1.1.1.1.1

18 Dave spins the spinner.

Which shape is impossible to spin?

Ⓕ

Ⓖ

Ⓗ

Ⓘ

MA.E.2.1.1.1.2   **Go On ▶**

19 Which does NOT show halves?

Ⓐ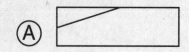

Ⓒ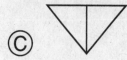

Ⓑ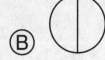

Ⓓ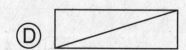

MA.A.1.1.3.1.2

20 Which shows how you can measure and compare?

Ⓕ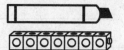

Ⓖ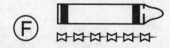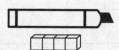

Ⓗ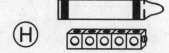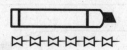

Ⓘ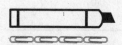

MA.B.2.1.2.1.1

21 Which fact belongs in this fact family?

$17 - 9 = 8$	$9 + 8 = 17$
$17 - 8 = 9$	?

Ⓐ $17 + 10 = 27$ Ⓒ $10 + 10 = 20$

Ⓑ $17 + 8 = 25$ Ⓓ $8 + 9 = 17$

MA.A.3.1.1.1.3   **Go On ▶**

Name_____

22 Look at the number chart.
Which does the chart show?

F Odd numbers are circled.

G Skip count by 2's.

H Skip count by 3's.

I Skip count by 5's.

1	2	3	4	5	6	7	8	9	10
11	12	13	14	15	16	17	18	19	20
21	22	23	24	25	26	27	28	29	30
31	32	33	34	35	36	37	38	39	40
41	42	43	44	45	46	47	48	49	50
51	52	53	54	55	56	57	58	59	60
61	62	63	64	65	66	67	68	69	70
71	72	73	74	75	76	77	78	79	80
81	82	83	84	85	86	87	88	89	90
91	92	93	94	95	96	97	98	99	100

MA.D.1.1.1.1.3

23 A bear slept 20 days.
Then it slept 30 days more.
How many days did the bear sleep?

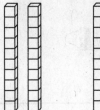 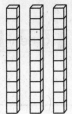

A 23

B 33

C 40

D 50

MA.A.3.1.3.1.1

24 About how many 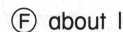 would you use
to balance a crayon?

F about 1

G about 2

H about 10

I about 50

STOP

MA.B.4.1.1.1.1

For 1 and 2, color to show order.

1.

first

second [yellow ▷ eighth [red ▷ tenth [blue ▷

2.

first

fourth [red ▷ sixth [blue ▷ ninth [green ▷

3. **Write the number words that match the dots. Use words from the box.**

zero	one	two
three	four	five
six	seven	eight
nine	ten	

_____ _____ _____ _____

© Harcourt

● **Draw a line from each word to its meaning.**

    1. dime   •

    2. doubles   •

    3. odd number •

    4. one fourth •

    5. estimate •

    • a number of objects that can be grouped into pairs and have one left over

    • one of four equal parts

    • two numbers in an addition fact that are the same

    • to tell how many you think there are

    • a coin worth 10 cents

● **Add. Then circle the doubles fact.**

    6. $5 + 6 =$ ____   7. $9 + 9 =$ ____

Circle the number you use to count back. Subtract.

    8. $12 - 3 =$ ____   9. $8 - 2 =$ ____

10. Use 4, 8 and 12. Write the four facts in the fact family.

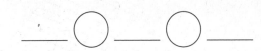

Number Sense, Concepts, and Operations

Draw a line from each object to the tool you would use to measure it.

1. How warm is the water?

   •   •

2. How much will the glass hold?

   •   •

3. How heavy is the book bag?

   •   •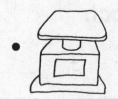

4. How long is the paper?

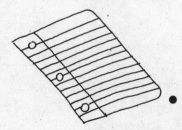   •   •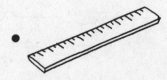

Measurement

I. Circle the pencil that is the longest.

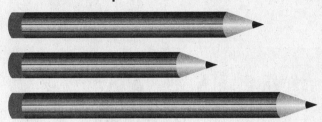

Draw the hour hand and the minute hand.

2.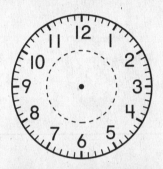

3.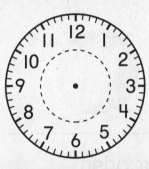

4.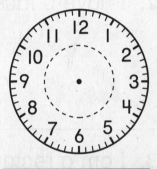

Use the calendar to answer the questions.

October						
Sunday	Monday	Tuesday	Wednesday	Thursday	Friday	Saturday
					1	2
3	4	5	6	7	8	9

5. What day of the week is October 4?
Color the day  blue .

6. On what day does the month begin?
Color the day red .

Measurement

Circle the solid that matches the sentence.

1. I am a cube.

2. I have 5 faces.

 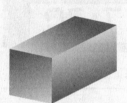

3. I am a rectangular prism.

 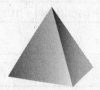

4. I have 6 faces.

5. I am a cylinder.

Geometry and Spatial Sense

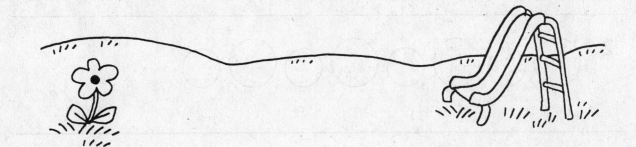

Follow the directions.

1. Draw a 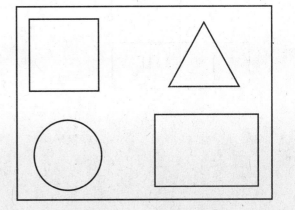 below the ⌒⌐.

2. Draw a 🕊 to the right of the ✈.

3. Draw a 🦋 over the 🌸.

4. Draw a 🌳 to the left of the ⌒⌐.

5. Trace each side of the rectangle.

6. Circle each vertex of the triangle.

Geometry and Spatial Sense

For 1, 2, and 3, circle the pattern unit.

1.

2.

3. □ ○ ○ □ □ ○ ○ □ ○ ○

Find the pattern. Then draw to continue it.

4. ▽ ▽ △ ▽ ▽ △ ▽ ▽ △ [   ]

Write the rule.

5.

Add ___	
5	8
3	6
1	4

6.

Subtract ___	
10	5
9	4
8	3

● **Draw a line from each symbol to its meaning.**

1. $<$ •   •is greater than

2. $=$ •   •is less than

3. $>$ •   •is equal to

Write $<$, $>$, or $=$ in the circle.

● 4.

2 tens 4 ones ◯ 30 + 4

5.

50 + 2 ◯ 5 tens 2 ones

6.

8 tens 6 ones ◯ 80 + 6

7.

70 + 9 ◯ 7 tens 8 ones

8.

9 tens 9 ones ◯ 80 + 9

9.

60 + 3 ◯ 6 tens 9 ones

Algebraic Thinking

Draw a line from each way to show information to its name.

1.

Fruits We Like		Total
♡apple	III	3
◡orange	⊬⊬⊬	5
🍇grapes	⊬⊬⊬ II	7

2.

Fruits We Like						
♡apple	🍎	🍎	🍎			
◡orange	◡	◡	◡	◡	◡	
🍇grapes	🍇	🍇	🍇	🍇	🍇	🍇	🍇

• picture graph

• tally table

• bar graph

3.   Fruits We Like

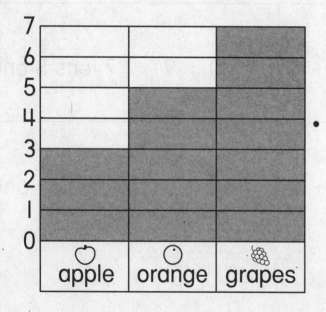

● **Mark an X to tell if pulling the cube from the bowl is certain or impossible.**

			Certain	Impossible
1.	▢	(bowl of cubes)		
2.	▢	(bowl of dark cubes)		
3.	■	(bowl of dark cubes)		

● **How likely is each shape to be pulled from the bowls? Draw ◯ or △ to tell.**

		More Likely	Less Likely
4.	(bowl of shapes)		
5.	(bowl of shapes)		

Data Analysis and Probability